Look at all the flowers! Each kind of flower has its own unique color, size, and shape.

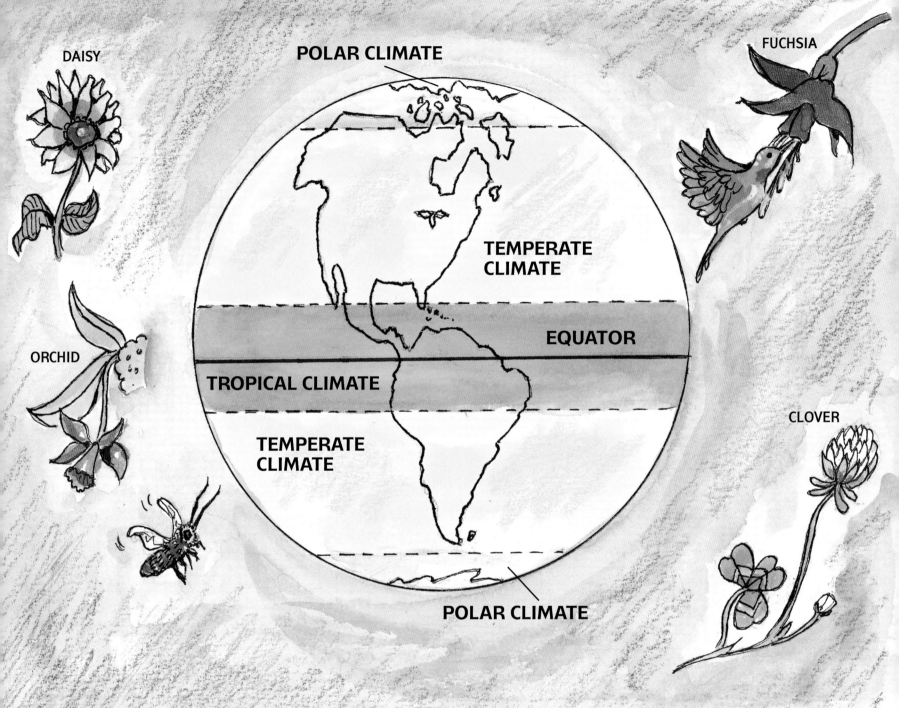

DAISY

POLAR CLIMATE

FUCHSIA

TEMPERATE CLIMATE

EQUATOR

ORCHID

TROPICAL CLIMATE

CLOVER

TEMPERATE CLIMATE

POLAR CLIMATE

Flowers grow in many places. Some flowers grow in temperate climates where there are four seasons. They grow only when it is not too cold. In tropical climates flowers can live longer because it is always warm.

PANSIES

FLOWERS

GAIL GIBBONS

Holiday House New York

To Veronica Walsh

Special thanks to Becky Sideman, horticulture specialist
at the University of New Hampshire in Durham.

Copyright © 2018 by Gail Gibbons
All Rights Reserved
HOLIDAY HOUSE is registered in the U.S. Patent
and Trademark Office
Printed and bound in November 2019
at Toppan Leefung, DongGuan City, China.
The artwork was created on watercolor paper with
black ink, watercolors, and colored pencil.
www.holidayhouse.com
First Edition
3 5 7 9 10 8 6 4 2

Library of Congress Cataloging-in-Publication Data
Names: Gibbons, Gail, author.
Title: Flowers / Gail Gibbons.
Description: First edition. | New York : Holiday House, [2019]
Identifiers: LCCN 2017022477 | ISBN 9780823437870 (hardcover)
Subjects: LCSH: Flowers—Juvenile literature.
Classification: LCC SB406.5 .G53 2019 | DDC 635.9—dc23 LC record available
at https://lccn.loc.gov/2017022477

ISBN: 978-0-8234-4537-0 (paperback)

BLEEDING HEART

WISTERIA

TEMPERATE CLIMATES

TEMPERATE CLIMATES have four seasons—spring, summer, fall, and winter. In winter it's too cold for flowers to survive.

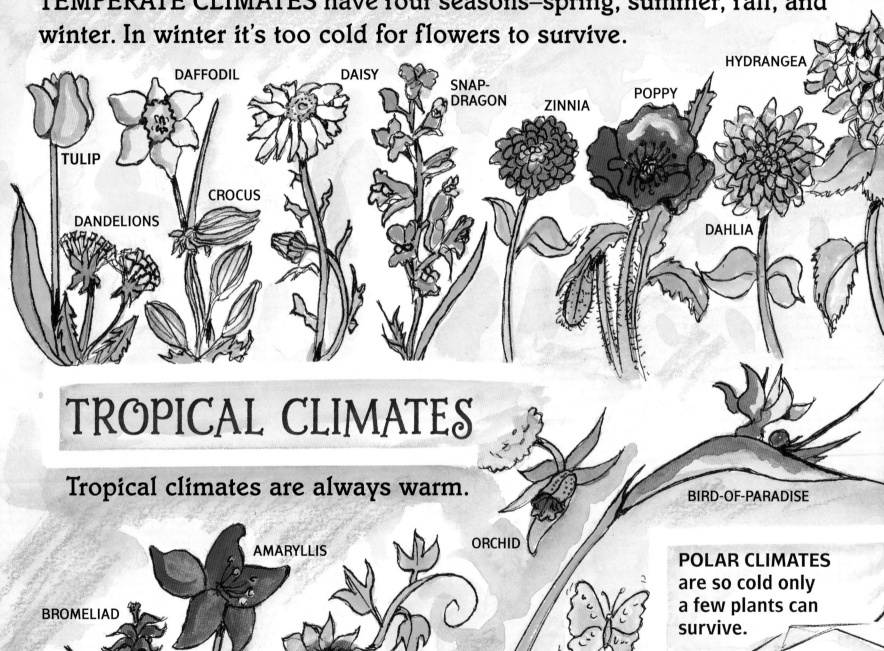

DAFFODIL

DAISY

SNAP-DRAGON

HYDRANGEA

ZINNIA

POPPY

TULIP

CROCUS

DANDELIONS

DAHLIA

TROPICAL CLIMATES

Tropical climates are always warm.

BIRD-OF-PARADISE

ORCHID

AMARYLLIS

POLAR CLIMATES are so cold only a few plants can survive.

BROMELIAD

PASSION-FLOWER

HIBISCUS

FLOWERS FROM SEEDS

Botanists are scientists who study plants.

FOXGLOVE

ZINNIA

PETUNIA

DAISY

SWEET PEA

COSMOS

DAISY

FLOWERS FROM BULBS

GLADIOLA

DAHLIA

LILY

HYACINTH

IRIS

Botanists group and identify flowers by how they grow.

SNOW PEA

6

FLOWERS FROM VINES

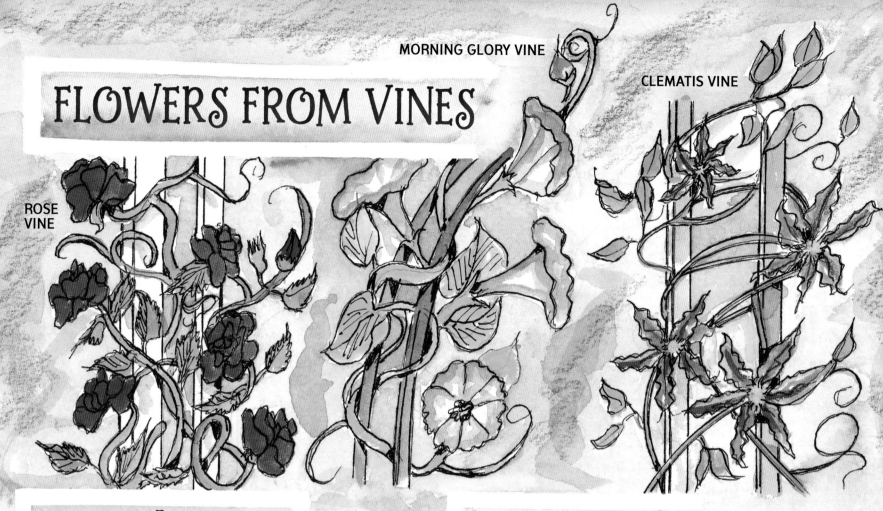

MORNING GLORY VINE

CLEMATIS VINE

ROSE VINE

FLOWERS ON BUSHES

AZALEA BUSH

BLACKBERRY BUSH

FORSYTHIA BUSH

FLOWERS ON TREES

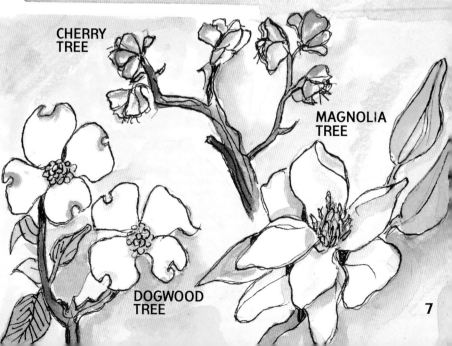

CHERRY TREE

MAGNOLIA TREE

DOGWOOD TREE

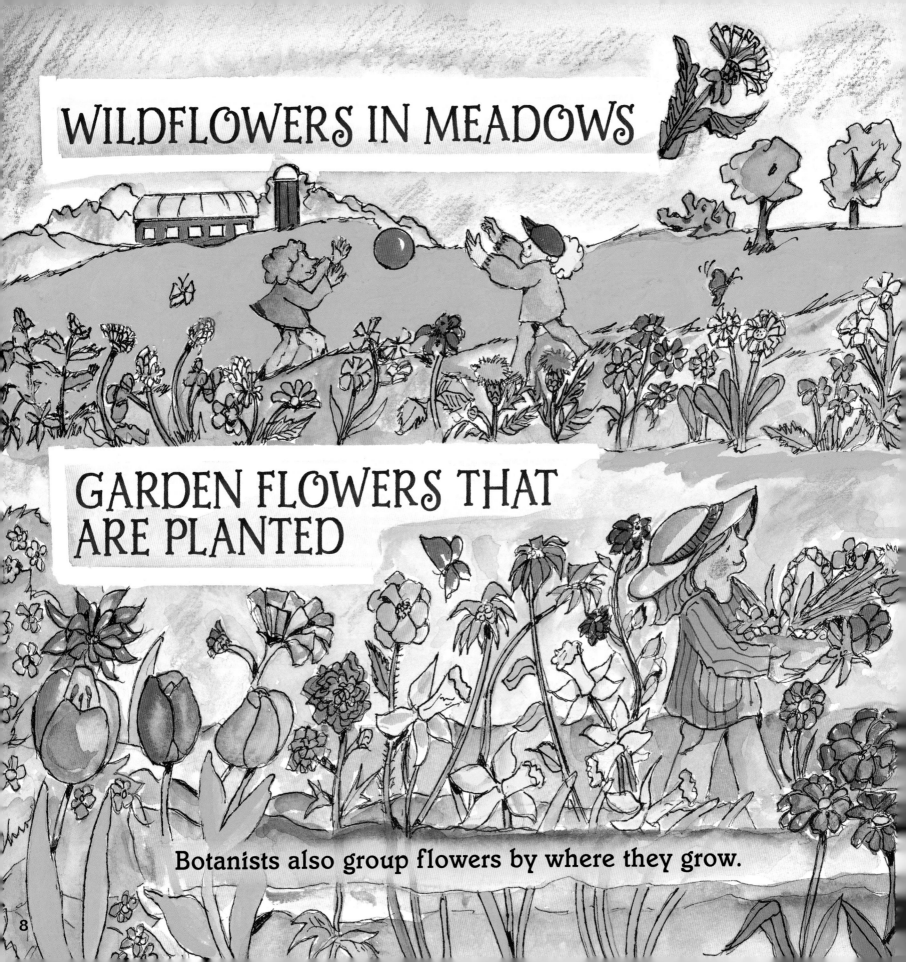

WILDFLOWERS IN MEADOWS

GARDEN FLOWERS THAT ARE PLANTED

Botanists also group flowers by where they grow.

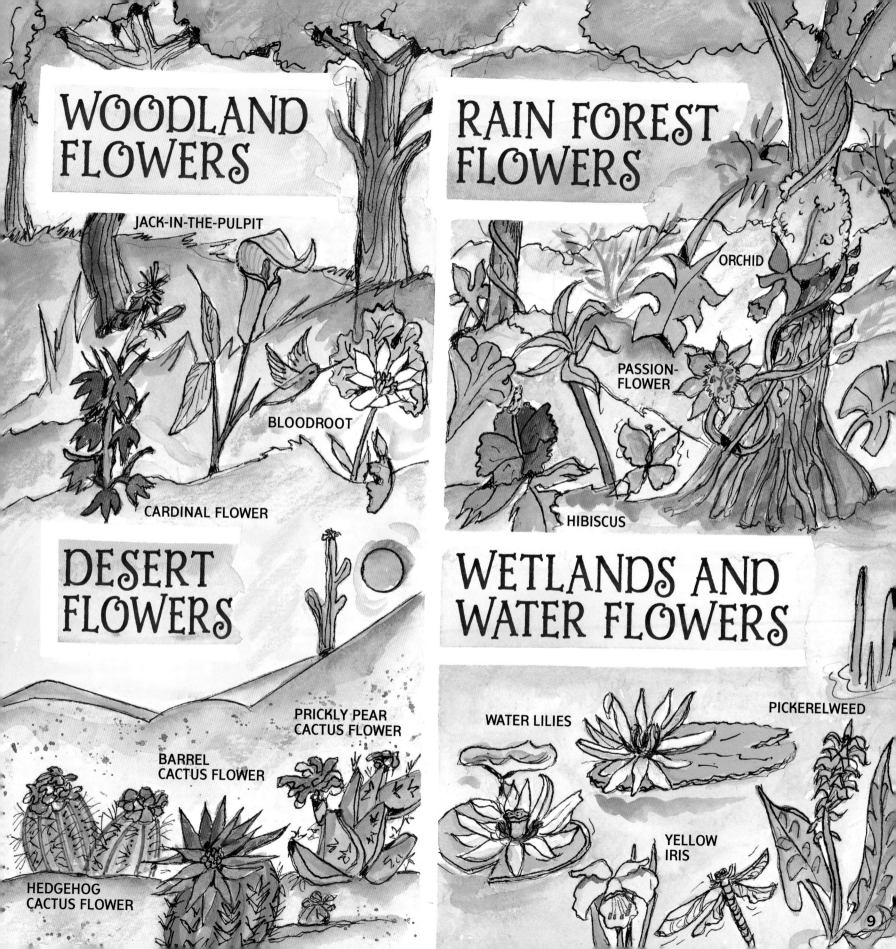

WOODLAND FLOWERS

JACK-IN-THE-PULPIT

BLOODROOT

CARDINAL FLOWER

RAIN FOREST FLOWERS

ORCHID

PASSION-FLOWER

HIBISCUS

DESERT FLOWERS

PRICKLY PEAR CACTUS FLOWER

BARREL CACTUS FLOWER

HEDGEHOG CACTUS FLOWER

WETLANDS AND WATER FLOWERS

PICKERELWEED

WATER LILIES

YELLOW IRIS

A FLOWER'S ENVIRONMENT

WATER LILY

Flowers need energy to grow. This comes from sunlight and carbon dioxide, a gas in the air.

SPECIES share specific characteristics.

DAISY

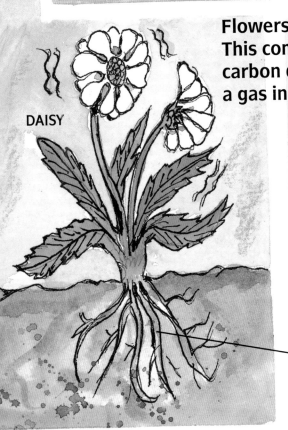

HEDGEHOG CACTUS

WATER and **NUTRIENTS** come up through **ROOTS.**

Some flowers grow best in SHADE.

Others need more SUNLIGHT.

SUNFLOWER

Some flowers CLOSE AT NIGHT for PROTECTION.

CROCUS

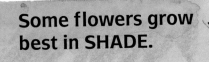

Each species of flower grows in the environment that is best for it.

LILY OF THE VALLEY

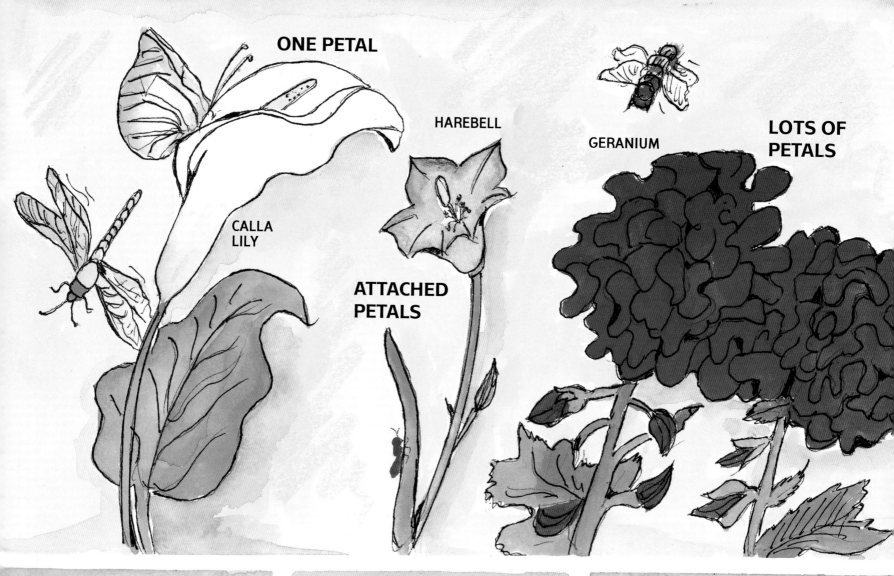

ONE PETAL

CALLA
LILY

HAREBELL

**ATTACHED
PETALS**

GERANIUM

**LOTS OF
PETALS**

**ONE FLOWER growing
from a stem**

VIOLET

**FLOWERS growing from
sides of a stem**

HYACINTH

**CLUSTERS OF FLOWERS
growing from a stem**

AZALEA

Flowers can look very different.

11

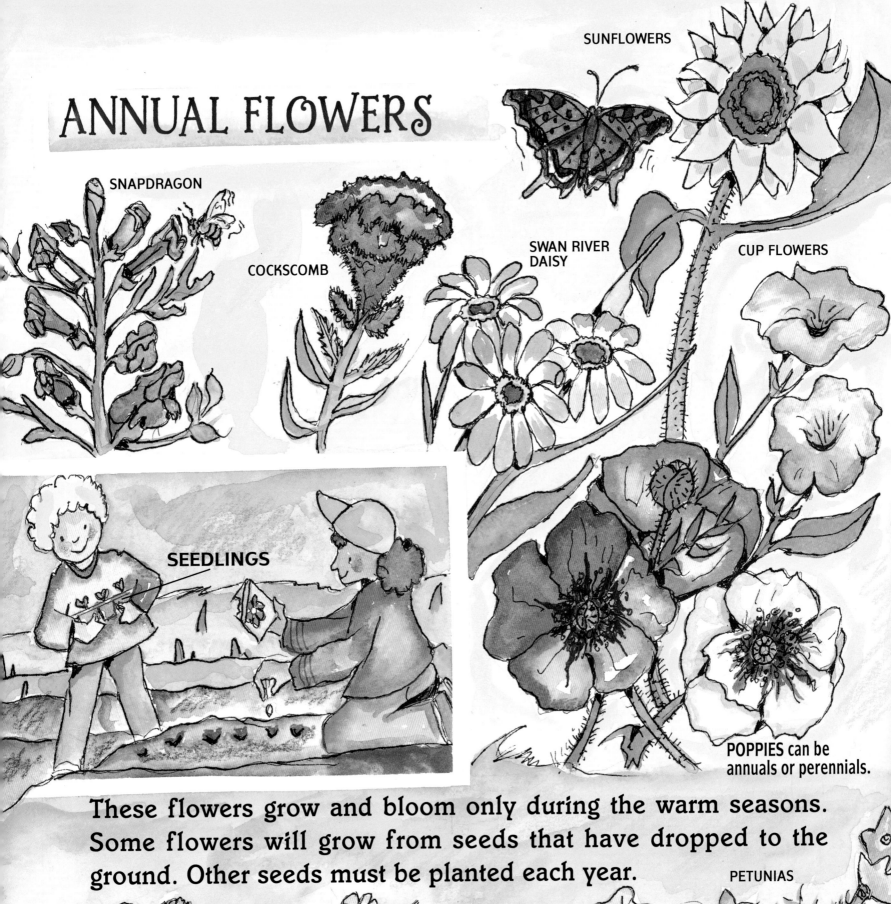

ANNUAL FLOWERS

SUNFLOWERS

SNAPDRAGON

COCKSCOMB

SWAN RIVER DAISY

CUP FLOWERS

SEEDLINGS

POPPIES can be annuals or perennials.

These flowers grow and bloom only during the warm seasons. Some flowers will grow from seeds that have dropped to the ground. Other seeds must be planted each year.

PETUNIAS

12

PERENNIAL FLOWERS

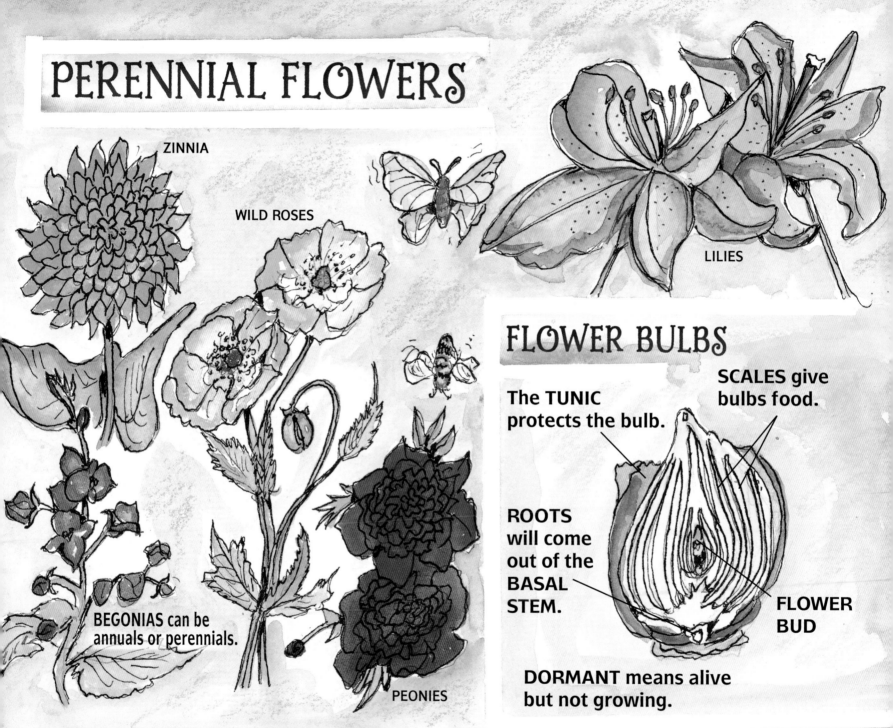

ZINNIA

WILD ROSES

LILIES

FLOWER BULBS

The **TUNIC** protects the bulb.

SCALES give bulbs food.

ROOTS will come out of the **BASAL STEM.**

FLOWER BUD

DORMANT means alive but not growing.

BEGONIAS can be annuals or perennials.

PEONIES

These flowers come up year after year. Some start as seeds. Others began as flower bulbs. Bulbs and the roots of perennials can stay alive underground through the winter. Flowering vines, bushes, and trees are dormant through the winter.

MARSH MARIGOLDS

THE PARTS OF A FLOWER

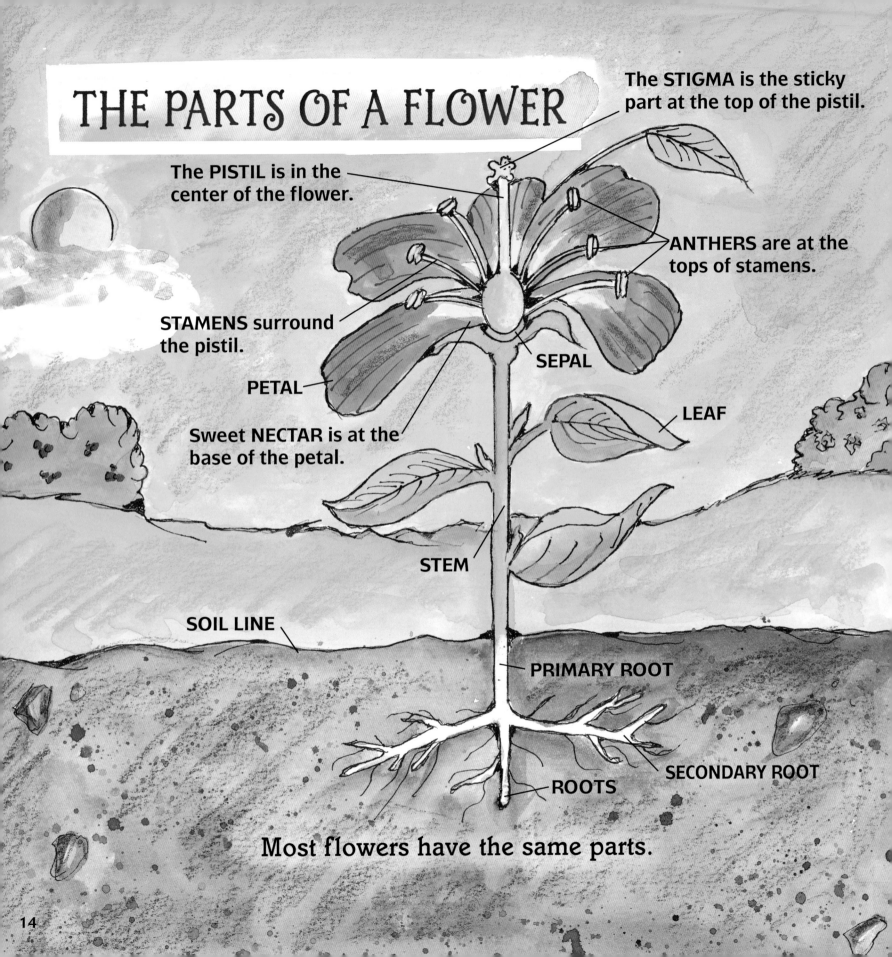

The STIGMA is the sticky part at the top of the pistil.

The PISTIL is in the center of the flower.

ANTHERS are at the tops of stamens.

STAMENS surround the pistil.

SEPAL

PETAL

LEAF

Sweet NECTAR is at the base of the petal.

STEM

SOIL LINE

PRIMARY ROOT

SECONDARY ROOT

ROOTS

Most flowers have the same parts.

POLLINATION

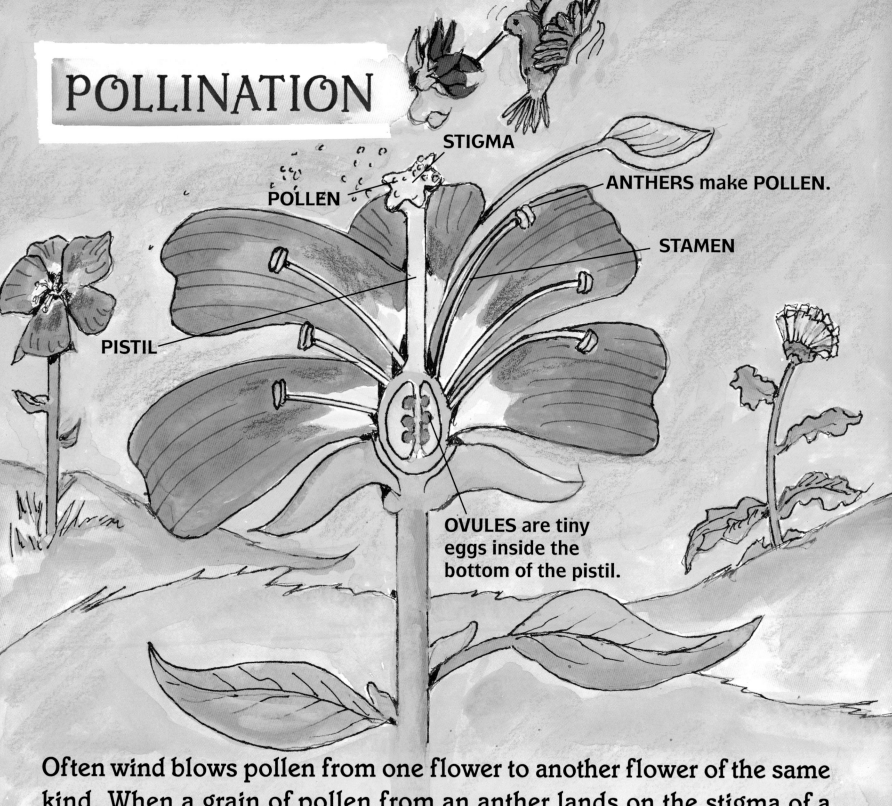

STIGMA

POLLEN

ANTHERS make POLLEN.

STAMEN

PISTIL

OVULES are tiny eggs inside the bottom of the pistil.

Often wind blows pollen from one flower to another flower of the same kind. When a grain of pollen from an anther lands on the stigma of a plant just like itself, pollination begins.

15

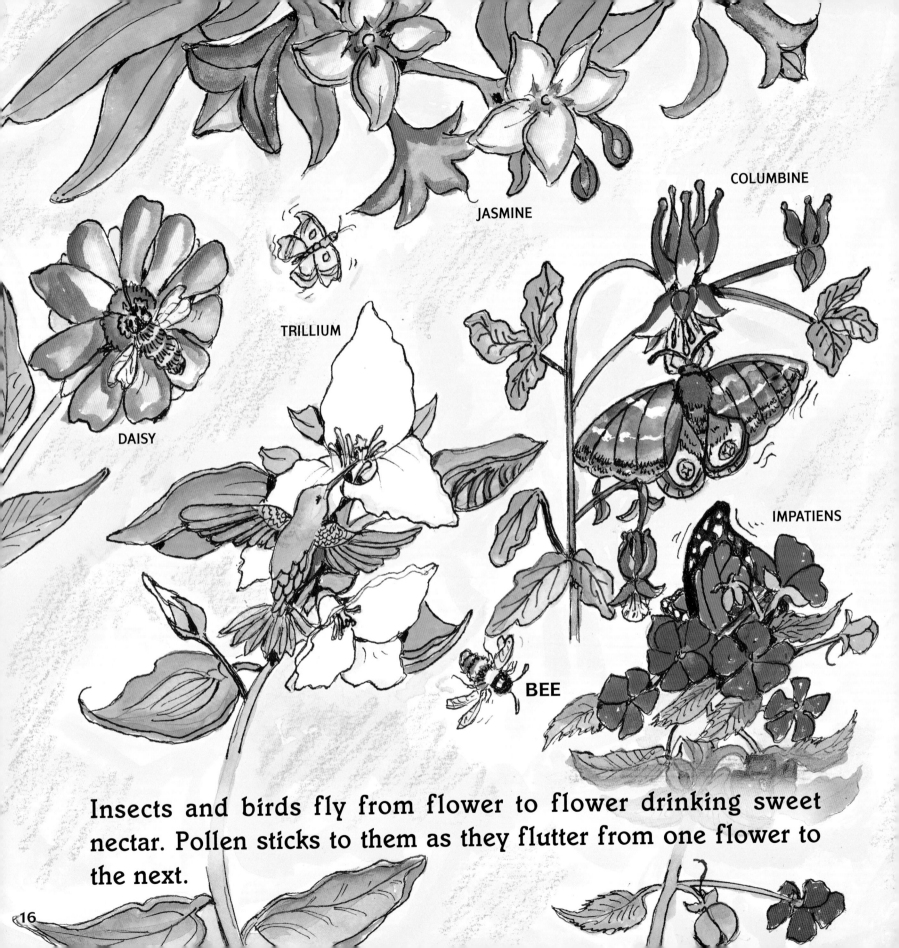

JASMINE

COLUMBINE

DAISY

TRILLIUM

IMPATIENS

BEE

Insects and birds fly from flower to flower drinking sweet nectar. Pollen sticks to them as they flutter from one flower to the next.

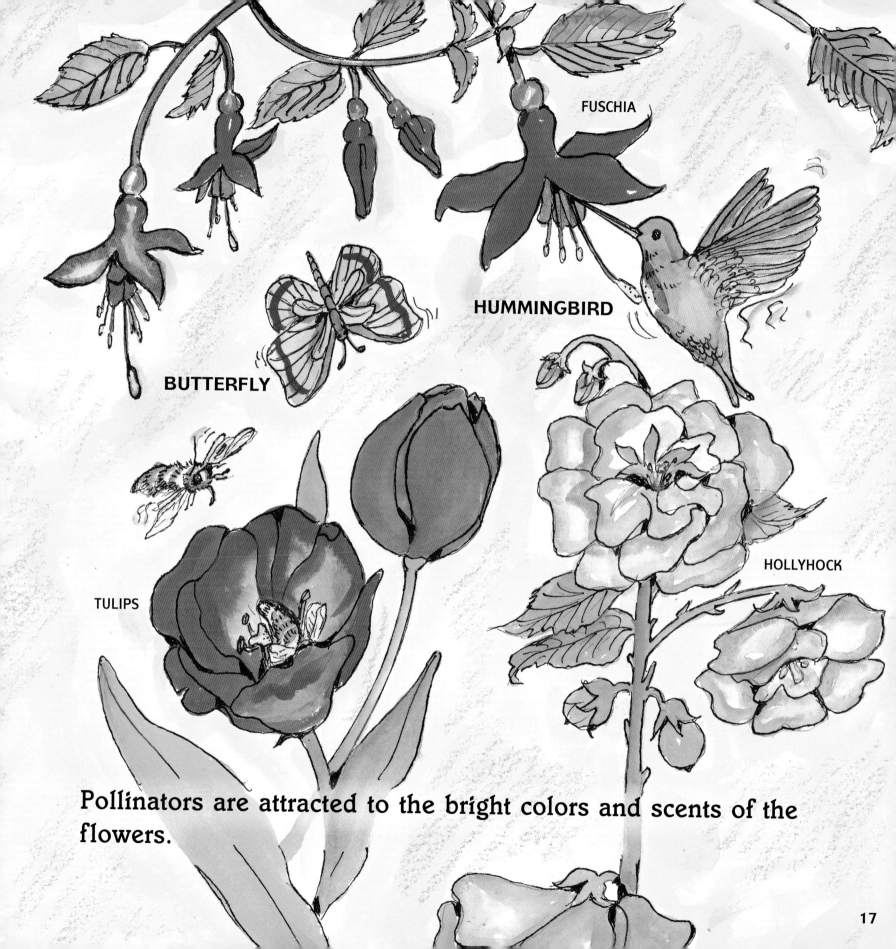

FUSCHIA

BUTTERFLY

HUMMINGBIRD

TULIPS

HOLLYHOCK

Pollinators are attracted to the bright colors and scents of the flowers.

STIGMA

POLLEN

PISTIL

NIGHTTIME POLLINATION

BAT

MOTHS

MOONFLOWER

ANGEL TRUMPET FLOWER

When pollinators fly to other flowers the pollen rubs off onto the sticky stigmas.

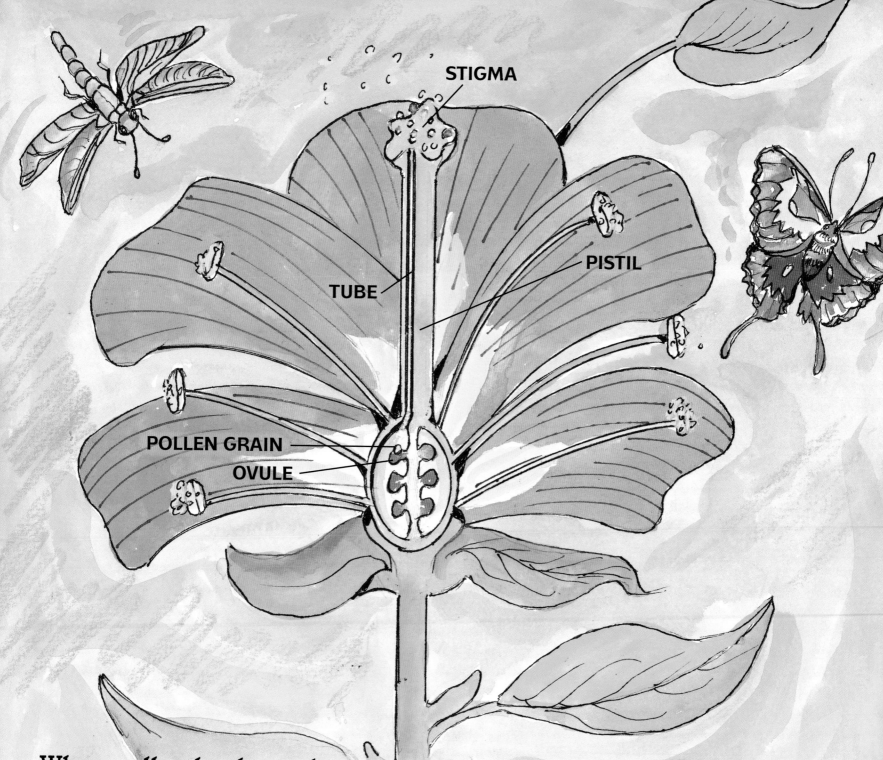

STIGMA

PISTIL

TUBE

POLLEN GRAIN

OVULE

When pollen lands on the stigma of the same kind of flower it came from, a long tube grows through the pistil. The pollen moves down the tube and fertilizes an ovule. A seed begins to grow.

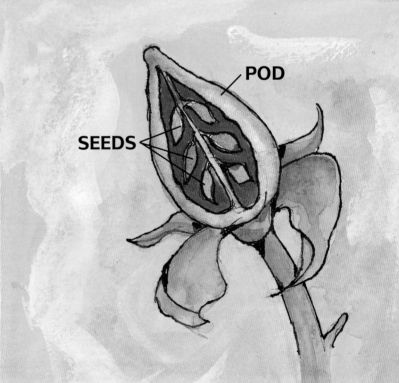

POD

SEEDS

Seeds grow inside the flower. As the flower dies the seeds get bigger. Often a pod grows around the seeds to protect them.

APPLE

SQUASH

Some seeds grow to become a fruit or vegetable we eat. Without pollination we wouldn't have fruits and vegetables.

HOW SEEDS TRAVEL

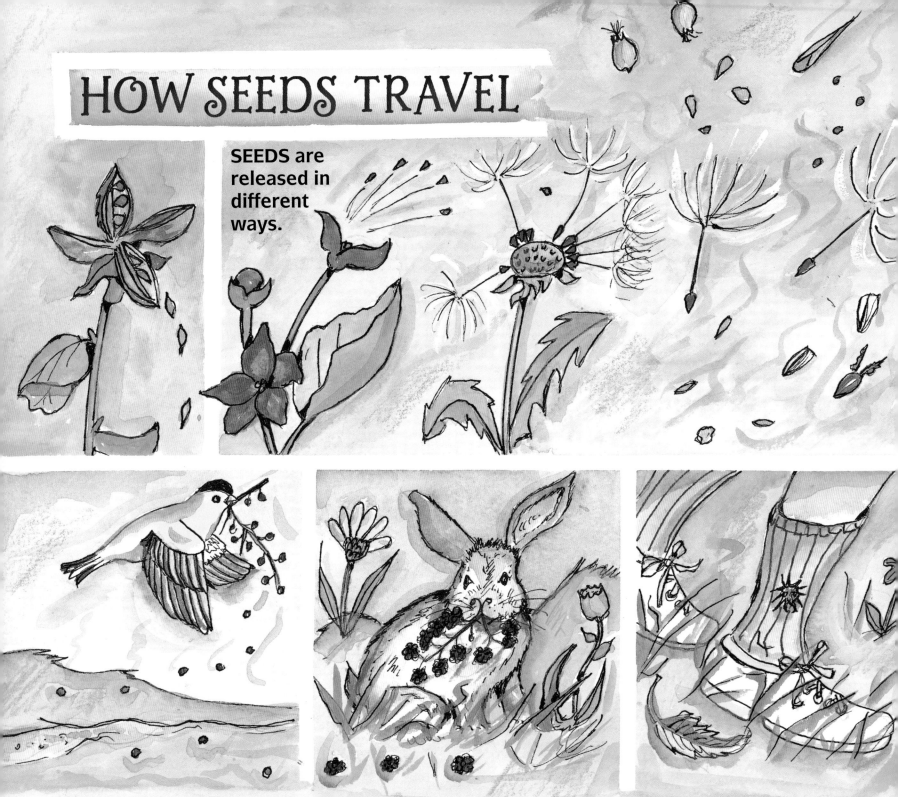

SEEDS are released in different ways.

When a flower pod ripens the seeds can fall out. The wind blows some seeds to the ground. Also, birds and animals move other seeds around.

The seeds travel to places where they can grow. Sometimes they have to wait until the next warm growing season.

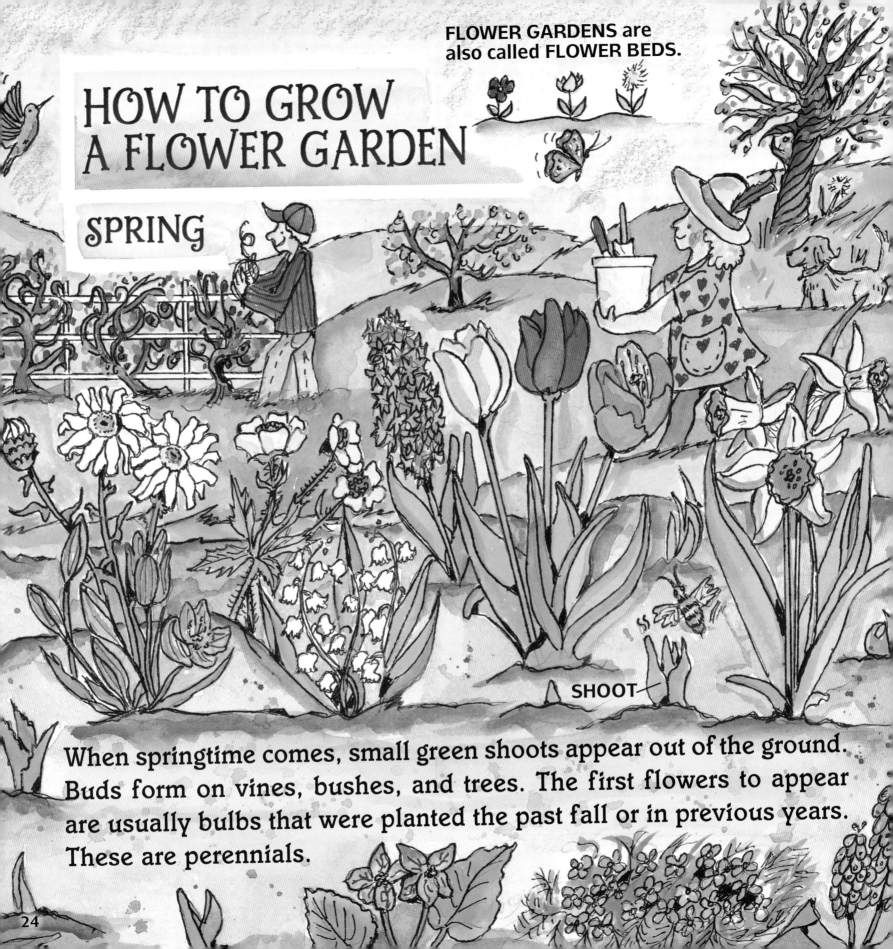

HOW TO GROW A FLOWER GARDEN

SPRING

SHOOT

When springtime comes, small green shoots appear out of the ground. Buds form on vines, bushes, and trees. The first flowers to appear are usually bulbs that were planted the past fall or in previous years. These are perennials.

ORGANIC FERTILIZERS and COMPOST are made up of broken-down leaves, grasses, and other waste, and are mixed into the soil.

SPADE

SHOVEL

RAKE

HOE

It's time to begin a new planting season. Clear the garden area and loosen up the soil. Plan the garden so each plant has enough room to grow.

25

SEEDLINGS

← SEEDS

SEED PACKETS

SUNFLOWER DAISY

ASTERS

LUPINE

LILIES

ASTERS ROSES BUSH

PERENNIAL PLANTS

Garden stores have seed packets and starter plants. Some people send away for seeds and plants. You can also choose full-grown plants to place in the ground or in containers.

VIOLET

WATERING
CAN

SUNFLOWER

HAND
TROWEL

HAND
CLAW
RAKE

WINDOW BOX

Follow the directions on how to plant your flower seeds. Place larger plants and seedlings into holes. Pack the soil around them. Be sure to water new plantings. It's fun to watch the springtime flowers grow.

SUMMER

SUNFLOWERS

WEEDING is important so flowers have room to grow.

SUPPORTS

SPRINKLER

HOSE

GLADIOLAS

FENCE

CLIMBING ROSES

TIGER LILIES

MORNING GLORIES

TRELLIS

WATERING CAN

DAHLIAS

ASTERS

The summer flowers blossom next.

BEGONIAS

28

MARIGOLDS

FALL

SQUASH

APPLES

MUMS

MARIGOLDS

ASTERS

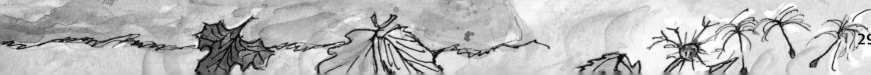

At last, when fall comes all the autumn flowers bloom. Some people plant next year's bulbs. Before it gets too cold, harvest fruits and vegetables and bring potted flowers inside to enjoy! Cover your flower beds to protect them from the cold.

29

GREENHOUSES protect seedlings and plants from harsh weather.

People like to pick flowers from their garden for family and friends.

OUR SCHOOL GARDEN

ANN JA MA DM CK

POTTING SOIL

DAISY

Some communities have gardens where people meet to plant flowers and care for them. Many schools plant flower gardens and grow flowers in their classrooms to study and enjoy.

For special occasions some people send flowers or go to florists to buy flowers for themselves or others. People love to enjoy flowers all year long!

FLOWER FACTS

BIRTHDAY FLOWERS

Find the flower for your birthday month.

The RAFFLESIA flower can be 3 feet (1 meter) wide and can weigh 15 pounds (7 kilograms).

The two biggest flowers are RAFFLESIA and the TITAN ARUM. Both flowers smell awful to attract insects for pollination.

The TITAN ARUM, also called the CORPSE FLOWER, can be 7 to 12 feet (2 to 3.5 meters) tall.

Some flower scents are used to make fragrant-smelling soaps, candles, perfumes, and other products.

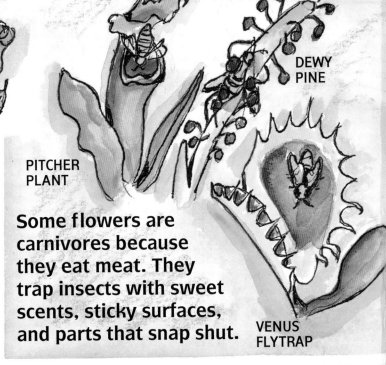

PITCHER PLANT

DEWY PINE

Some flowers are carnivores because they eat meat. They trap insects with sweet scents, sticky surfaces, and parts that snap shut.

VENUS FLYTRAP

CAULIFLOWER

BROCCOLI

We eat some flowers such as cauliflower, broccoli, and Brussels sprouts.

BRUSSELS SPROUTS

BELLFLOWER

Flowers follow the movement of the sun during the day.

Birthday Flowers

JANUARY — CARNATION

FEBRUARY — VIOLET

MARCH — DAFFODIL

APRIL — DAISY

MAY — LILY OF THE VALLEY

JUNE — ROSE

JULY — WATER LILY

AUGUST — GLADIOLA

SEPTEMBER — ASTER

OCTOBER — MARIGOLD

NOVEMBER — CHRYSANTHEMUM

DECEMBER — HOLLY